Berichte zu Tierarzneimitteln 2009

Gesundheitliche Bewertung von pharmakologisch wirksamen Substanzen – Lebensmittelsicherheit von Rückständen von Tierarzneimitteln

Target Animal Safety for Veterinary Pharmaceutical Products (VICH GL 43)

Beurteilung und Überwachung der Resistenzsituation bei und nach der Zulassung von Tierarzneimitteln

Inhaltsverzeichnis

1. Stefan Scheid
   Gesundheitliche Bewertung von pharmakologisch wirksamen Substanzen – Lebensmittelsicherheit von Rückständen von Tierarzneimitteln.. 5

2. Gesine Hahn
   Target Animal Safety for Veterinary Pharmaceutical Products (VICH GL 43) 58

3. Christine Schwarz
   Beurteilung und Überwachung der Resistenzsituation bei und nach der Zulassung von Tierarzneimitteln 70

1 Gesundheitliche Bewertung von pharmakologisch wirksamen Substanzen – Lebensmittelsicherheit von Rückständen von Tierarzneimitteln

Stefan Scheid

Bundesamt für Verbraucherschutz und Lebensmittelsicherheit (BVL), Berlin

Korrespondenz an: Dr. S. Scheid, BVL, Abt. 3 Tierarzneimittel, Ref. 305 Rückstände pharmakologisch wirksamer Stoffe, Mauerstraße 39–42, 10117 Berlin,
Tel.: 030 18 444 30500, e-mail: Stefan.Scheid@bvl.bund.de

BVL Referat 305 - Aufgabenschwerpunkte

Lebensmittelsicherheit (im Zulassungsverfahren)
- Höchstmengen (MRL) für Rückstände von Tierarzneimitteln
- Festlegung von Wartezeiten
- Prüfung von analytischen Kontrollmethoden ("desk review")

Anwendersicherheit
- „User Safety" - Unbedenklichkeit für den Anwender

Zieltiersicherheit
- Toxikologische Unbedenklichkeit für das Zieltier

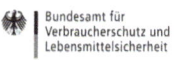

Rechtlicher Kontext

- **VO (EG) Nr. 470/2009**
  Festlegung von Rückstandshöchstmengen (Maximum Residue Limits, MRL) für pharmakologisch wirksame Stoffe in Lebensmitteln vom tierischen Ursprungs

- **Richtlinie 2001/82/EG und VO (EG) Nr. 726/2004**
  Europäische Zulassungsverfahren für Tierarzneimittel (dezentralisierte/zentralisierte Verfahren)

- **Arzneimittelgesetz**
  Nationale Zulassungsverfahren

08.08.2009 — St.Sch.• 2009 4

Festlegung von Rückstandshöchstmengen (Maximum Residue Limits MRL)

08.08.2009 — St.Sch.• 2009 5

MRL - Basic Law

16.6.2009 EN Official Journal of the European Union L 152/11

REGULATION (EC) No 470/2009 OF THE EUROPEAN PARLIAMENT AND OF THE COUNCIL

of 6 May 2009

laying down Community procedures for the establishment of residue limits of pharmacologically active substances in foodstuffs of animal origin, repealing Council Regulation (EEC) No 2377/90 and amending Directive 2001/82/EC of the European Parliament and of the Council and Regulation (EC) No 726/2004 of the European Parliament and of the Council

(Text with EEA relevance)

eur-lex.europa.eu/LexUriServ/LexUriServ.do?uri=OJ:L:2009:152:0011:0022:EN:PDF

MRL - What are they for?

Maximum residue limits are the points of reference for:

- ✓ the establishment of **withdrawal periods** for veterinary medicinal products to be used in food-producing animals (Dir.2001/82/EC, CR 726/2004)

- ✓ the **control of residues** in food of animal origin in the Member States and at border inspection posts (Dir.96/23/EC)

Structure of the Regulation
- Types of Reference Values

Title I   General Provisions - Scope and Definitions

➢ Title II   Maximum Residue Limits (MRL)

   Chapter I   Risk Assessment and risk management

   •Section 1 substances for use in VMPs
   •Section 2 substances not intended for use in VMPs

   Chapter II   Classification of pharmacologically active substances

➢ Title III   Reference points for action

Maximum Residue Limits
- Opinion by the EMEA

➢ Substances: Intended for use in VMP
➢ Action by: Pharmaceutical company seeking MA

Article 3

- Substance for use in VMP for food-producing animals in the EU shall be subject to an opinion of the EMEA on the maximum residue limit

- Exception: A Codex Alimentarius MRL is available (without objection from EU Community Delegation)

Article 4

- EMEA to carry out scientific risk assessment and draw up risk management recommendations

Maximum Residue Limits
- Opinion by the EMEA

- **Substances:** <u>Not</u> intended for use in VMP
- **Action by:** Commission or MS request for MRL

Article 9

- <u>Substance authorised as VMP in a third country</u> - no application pursuant to Article 3
- <u>Substance is included in VMP to be used pursuant to Article 11 of Directive 2001/82/EC (cascade)</u> - no application pursuant to Article 3

Article 10

- <u>Substances contained in biocidal products</u> used in animal husbandry, as defined in Directive 98/8/EC. Classification of substances in accordance with Article 14

Classification of Maximum Residue Limits

4 Classes of substances
(Comparable to Annexes 1, 3, 2 and 4 of Regulation 2377/90)

1. A maximum residue limit
   - On the base of opinion of EMEA
   - Vote by Community in Codex Alimentarius

2. A provisional maximum residue limit
   - Incomplete data, but no grounds for supposing a relevant risk

3. The absence of a maximum residue limit
   - Not necessary for protection of human health

4. A prohibition on the administration of a substance
   - Where any residue may constitute a hazard
   - where no final conclusion concerning human health with regard to residues of a substance can be drawn.

Who are the parties involved/concerned?

- Regulators
  - European Medicines Agency (EMEA)
  - European Food Safety Authority, EFSA
    European Commission
  - Member States of the EU (27)/National Competent Authorities

- Pharmaceutical Industry

- Veterinarian and Producers

- Consumer

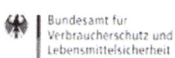

"Maximum Residue Limits"
– Procedure for VMP

Applicant } Scientific Studies Hazard & Exposure

EMEA } Risk Assessment & Management Options

COMMISSION } Regulation

Article 14(2) (a - d)

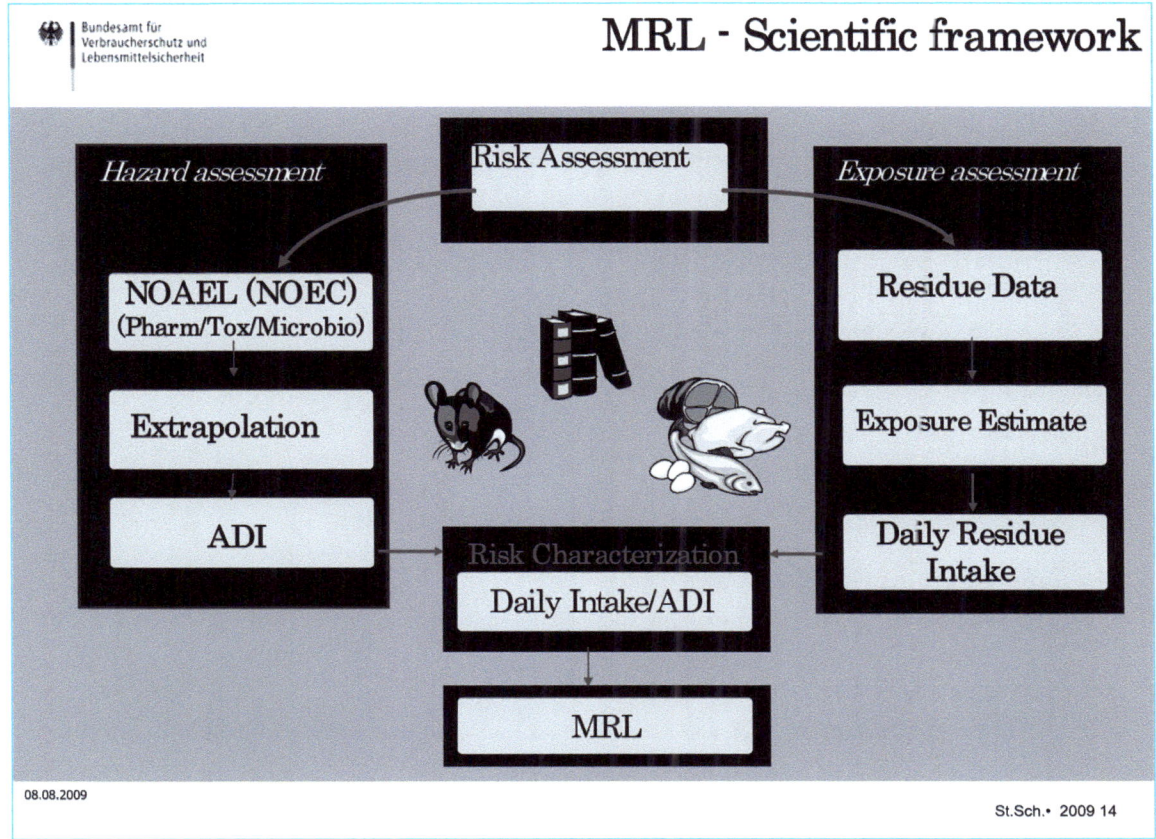

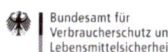

Maximum Residue Limits
– Hazard Assessment

Different types of Hazards and ADIs

Pharmacol. hazard → $ADI_{PHARM.}$
Toxicolog. hazard → $ADI_{TOXICOL.}$
Microbiolog. hazard → $ADI_{MICROBIOL.}$

} Overall ADI

The concept of ARfD is not used!

Maximum Residue Limits
– Hazard Assessment

Steps to be taken:

1. Review entire database for pharmacological and toxicological effects and identify the critical effects/endpoints

2. Determine for each critical endpoint the NO(A)EL (LO(A)EL) as the <u>highest</u> dose at which there is no statistically significant difference between the effects seen in test animals and controls experiment. Make use of dose-response relationship, where possible

3. Identify the (overall) <u>NO(A)EL</u> for the most sensitive effect in the most sensitive species (lowest NO(A)EL) as the basis for the ADI

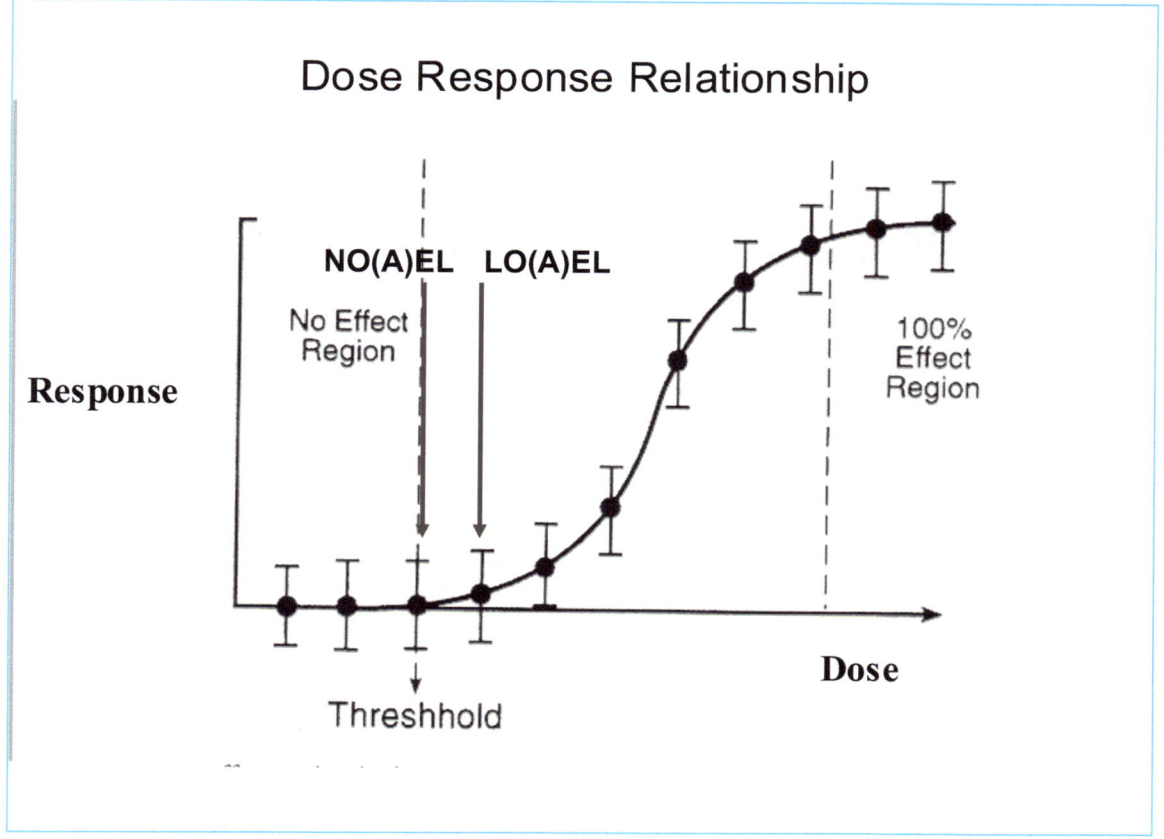

Maximum Residue Limits – Hazard Assessment

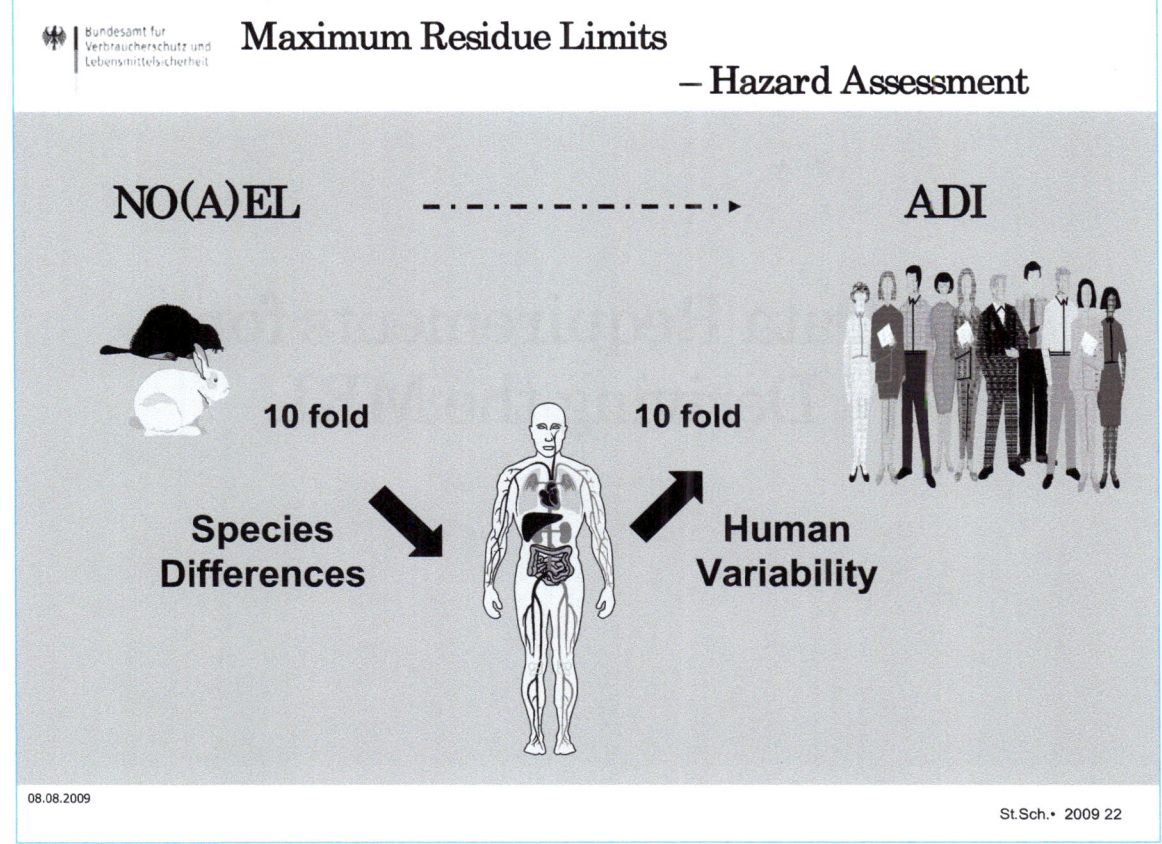

Maximum Residue Limits – Hazard Assessment

A few words on the Microbiological ADI

Hazards related to use of antibiotics:

- ✓ Disruption of colonisation barrier function of intestinal flora

- ✓ Increasing the population of resistant bacteria (acquisition of resistance/increase in the proportion of less sensitive organisms)

- ✓ Determination of NOEC for both endpoints using MIC data and other information

- ✓ The NOEC is POD for a "microbiological" ADI in humans (safe dose to cover microbiological hazards)

Data Requirements for Deriving the MRL

Data Requirements

Five types of information needed:

- → Acceptable Daily Intake (ADI)
- → Comparative metabolism studies
- → Total residue depletion studies in target species
- → Specific marker residue depletion studies in target species
- → Analytical methods suitable for residue control *(validation criteria discussed later)*

→ Acceptable Daily Intake (ADI

- Estimate of the residue that can be ingested <u>daily over a lifetime</u> without any appreciable health risk (e.g., Glossary of "Volume 8")

- Central health based guidance value in the ("chronic") exposure assessment of veterinary drug residues (and assessments in other areas of food safety)

- MRLs are established such that the maximum possible intake of residues does not exceed the ADI (when VMP are used according to label instructions)

→ Total Residues

- Residue definition includes all residues, active substance(s), also metabolites, degradation products

<u>And:</u>

- All drug derived residues are assumed to have the same hazard potential as parent compound (in the absence of evidence to the contrary)

<u>Therefore:</u>

- A basic requirement is the conduct of studies that demonstrate the total drug derived residues

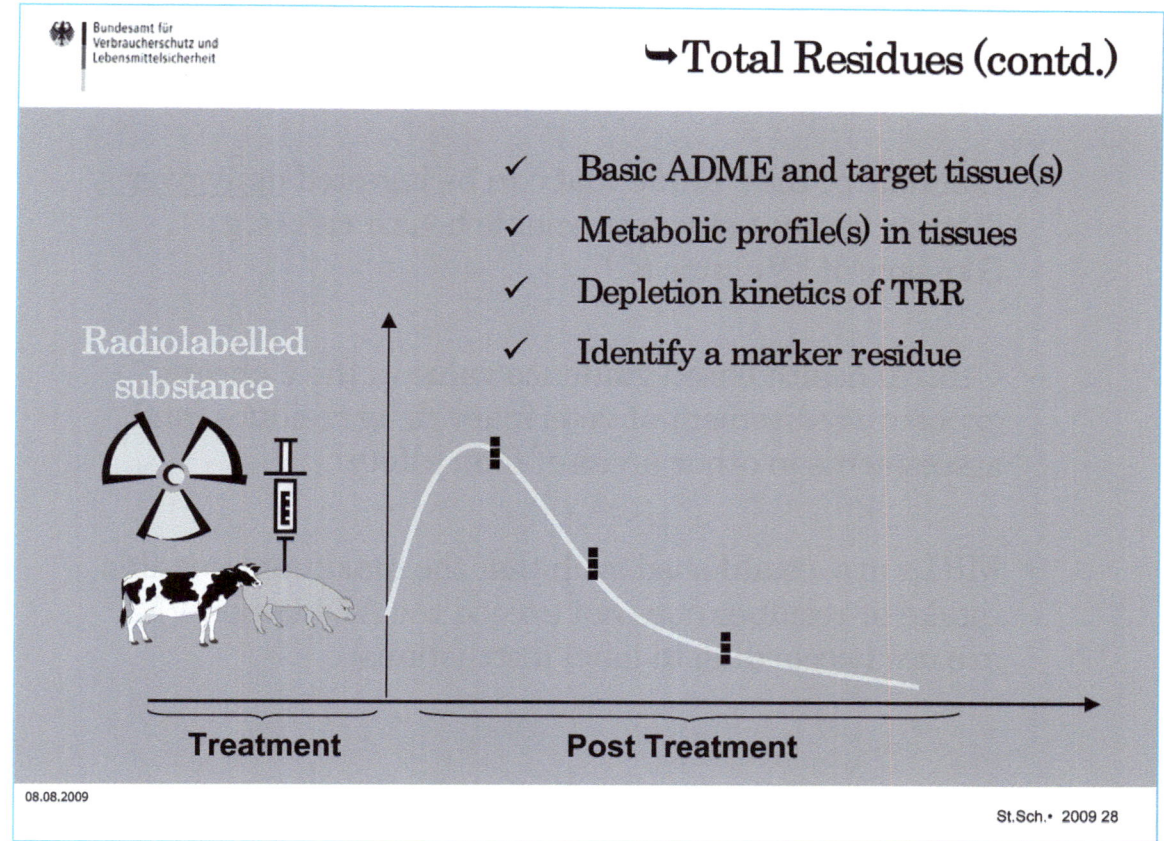

Data requirements - Checklist

- ✓ ADI
- ✓ Established comparability of metabolism target vs tox species
- ✓ Total residue data
- ✓ Marker residue
- ✓ Marker residue depletion data

Exposure Assessment

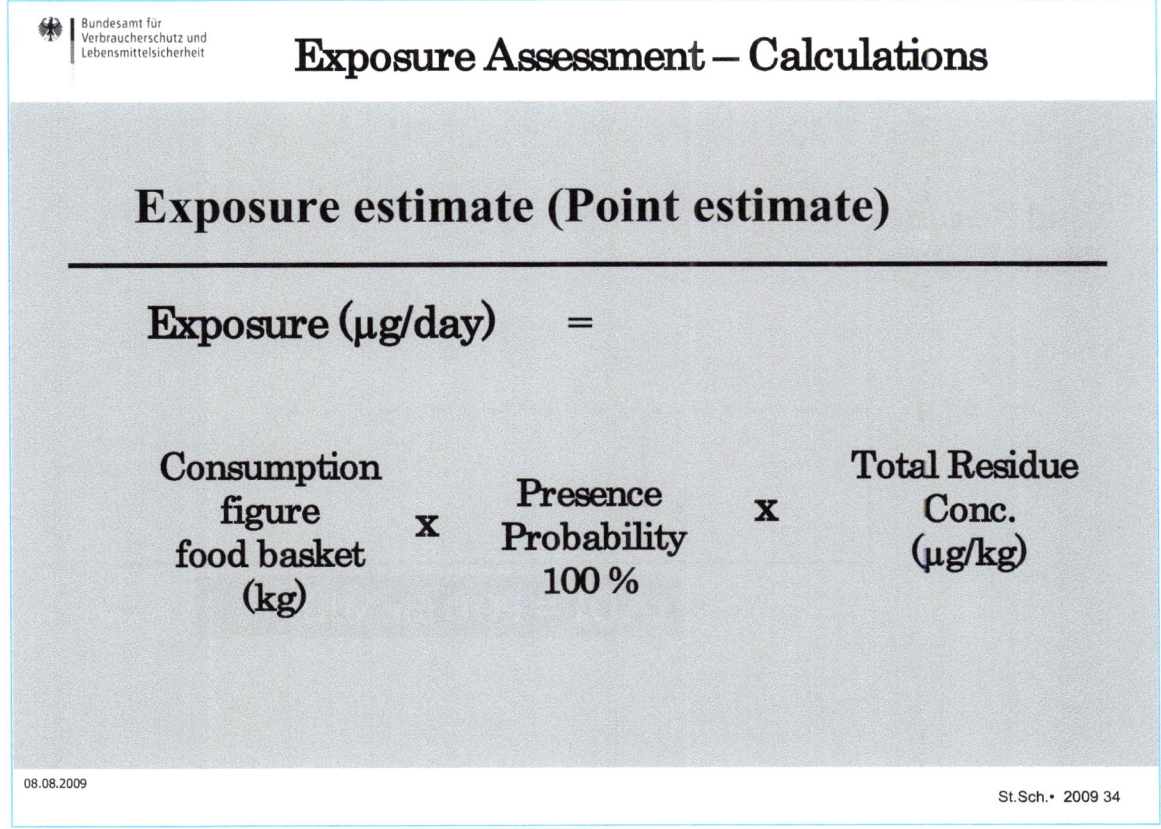

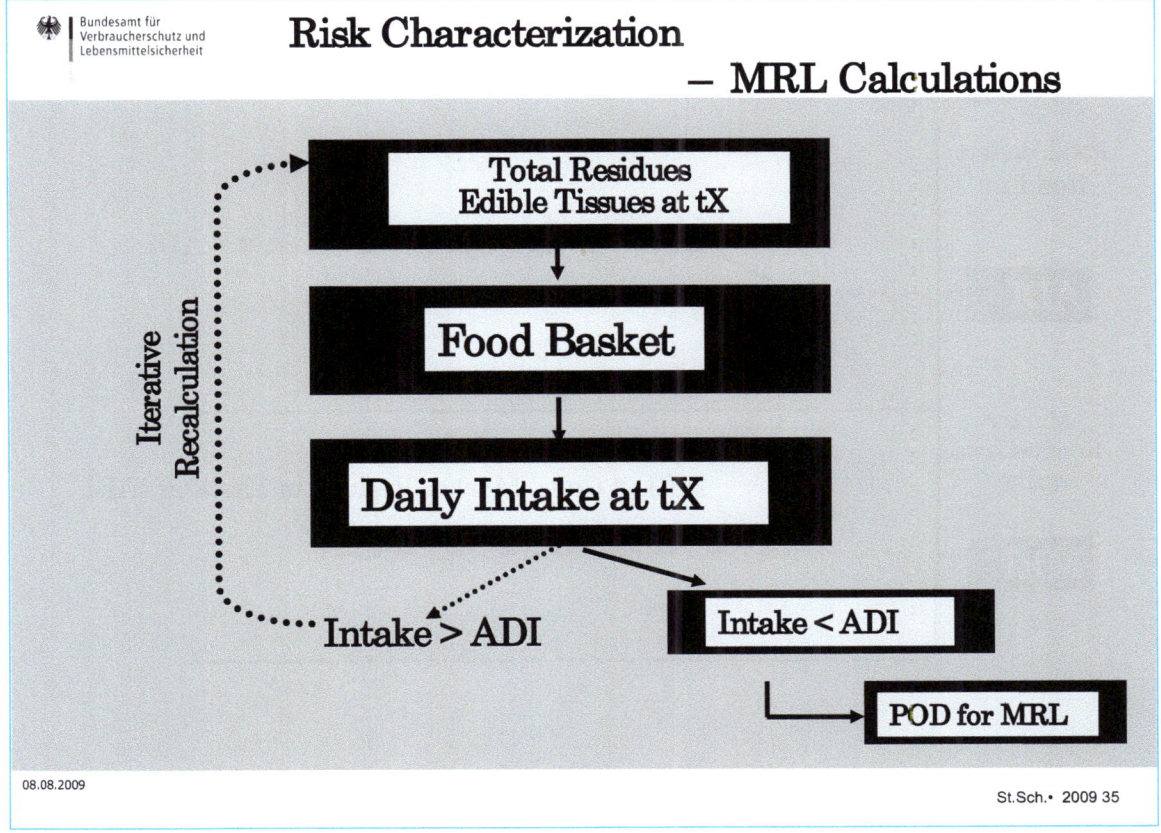

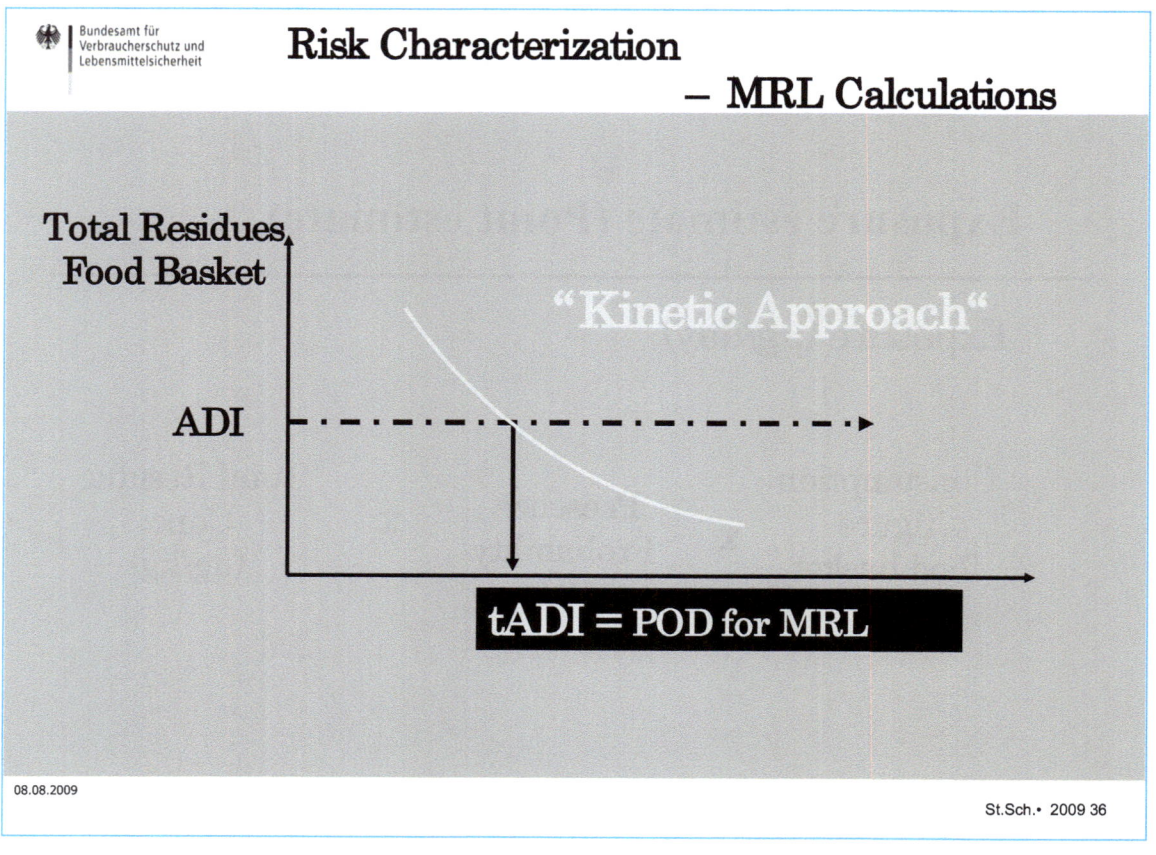
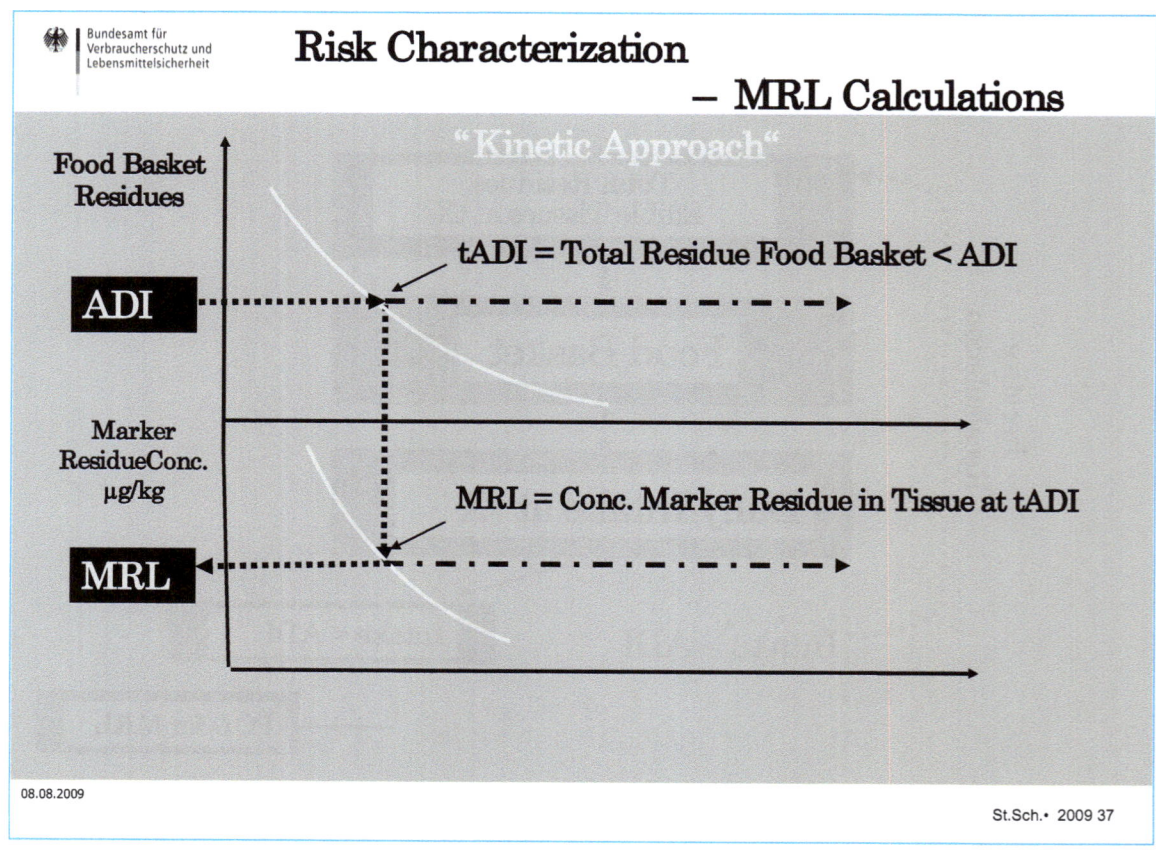

For which tissues are MRL calculated?

Target tissues for MRLs all edible tissues/products from treated animals:

- Liver
- Kidney
- Muscle (Muscle/Skin)
- Fat (Fat/Skin)
- (Inj. Site)

but also:

- milk
- eggs
- honey

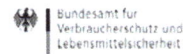

How is the ADI divided between target tissues?

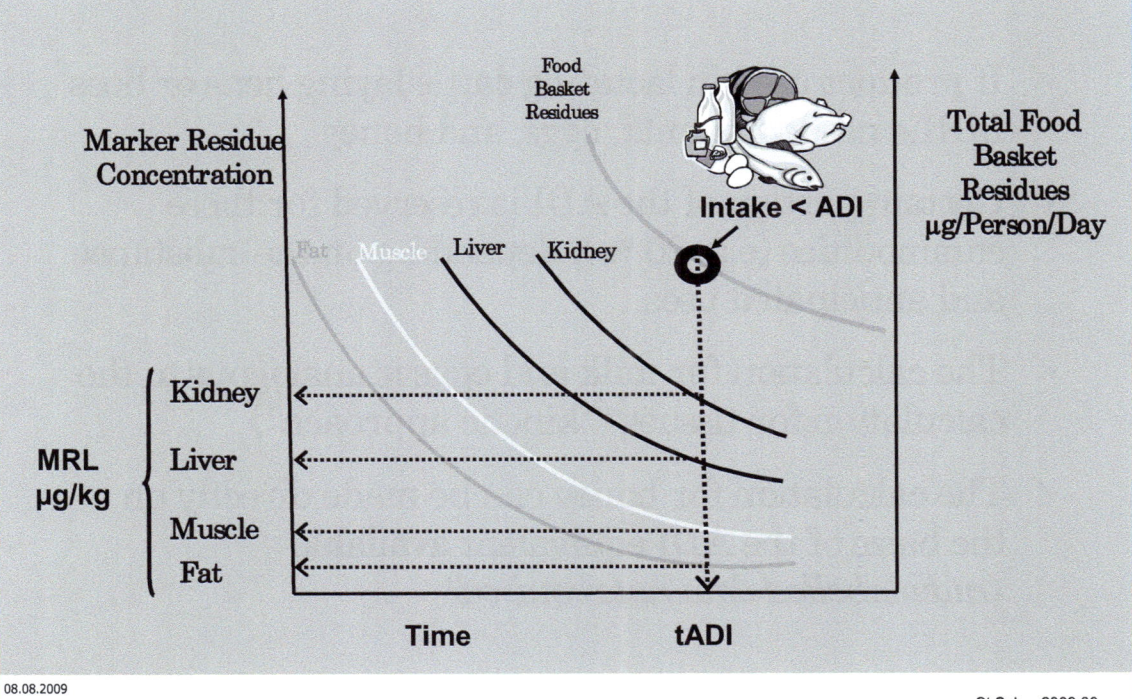

The Tissue Distribution Concept (TDC)

Why is it used?

- ✓ Tissue distribution concept means that MRLs are allocated following the actual residue distribution observed in the different in target tissues (tissues with highest residues get the highest MRL, and vice versa)

- ✓ If this concept is followed, each target tissue MRL is a monitor for depletion of residues from all other edible tissues to below the ADI and can serve as tissue for control of residues

- ✓ In addition, the withdrawal periods for each edible tissue will approximately be the same

MRL in milk, eggs, honey

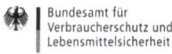

- ✓ If product used in lactating cattle laying hens or bees MRLs needed in milk, eggs, and honey

- ✓ Certain portion of the ADI is reserved for these commodities (e.g. 20 %), depending on the substance and anticipated uses

- ✓ The calculation for milk and eggs is analogous to the calculation for tissues ("kinetic approach")

- ✓ The calculation for honey can be made directly on the basis of the ADI equivalent available *(radiolabelled data not required)*

Dual use and aggregate exposure

- CR (EC) No 470/2009: "The scientific risk assessment shall concern [....] residues that occur in <u>food of plant origin or that come from the environment</u>"

- For so-called dual-use substances, i.e., substance that are used as pesticides or biocides or that might occur in drinking water " aggregate exposure" will be considered

- The combined exposure from these sources will be subtracted from the ADI, thus lowering the portion of the ADI available for veterinary uses

MRL calculation - Summary

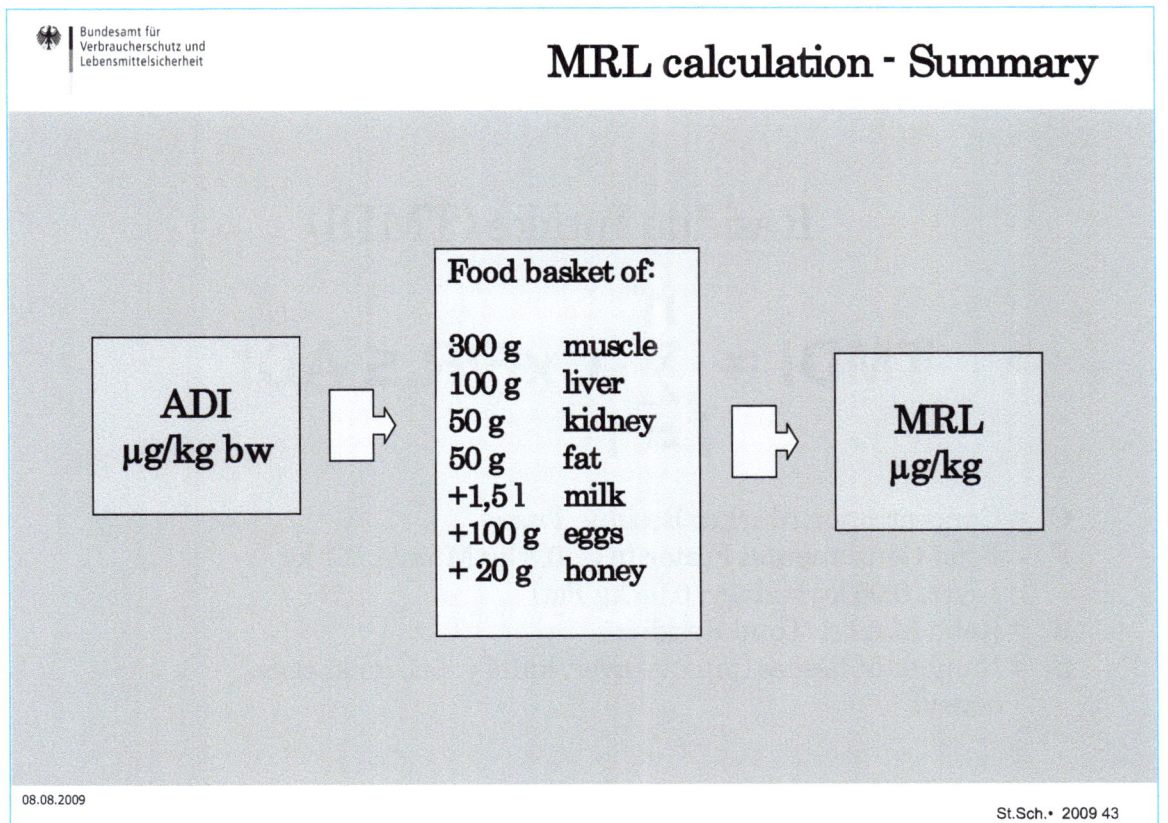

MRL calculation - Summary

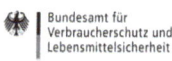

Step 1 Determine the time point when total residue in food basket are below ADI (t_{ADI})

Step 2 Allocate a portion of the ADI to each target tissues (ideally based on TDC concept)

Step 2 Choose a Marker Residue suitable for residue monitoring and estimate an MRL in each target tissue

Step 4 Check if the estimated MRLs are fully in compliance with the ADI, i.e. if the TMDI (based on food basket consumption figures and ratio marker/total residues) is ≤ ADI

MRL calculation - Summary

Residue Intake (TMDI)

$$TMDI = \sum_{i=1}^{n} CxF/R \leq ADI$$

C = Concentration Marker Residue Tissue
F = Food Consumption Factor (e.g., 0.3 kg Muscle, 0.1 kg Liver, 0.05 kg Kidney, 0.05 kg Fat)
R = Ratio Marker/Total Residue
n = Number of tissues (muscle, liver, kidney, fat, milk, eggs, honey)

Extrapolation of MRLs

CR (EC) No 470/2009

" maximum residue limits established for a pharmacologically active substance in a particular foodstuff [*may be used*] for another foodstuff derived from the same species, or maximum residue limits established for a pharmacologically active substance in one or more species for other species"

Extrapolation of MRLs

……and Volume 8 allows MRLs to be extrapolated within classes of animals

Species for which MRLs have been set	Extrapolations to
Major ruminant	All ruminants
Major ruminant milk	All ruminant milk
Major monogastric mammal	Extrapolation to all monogastric mammals
Chicken and eggs	Poultry and poultry eggs
Salmonidae	All fin fish
Either a major ruminant or a major monogastric mammal	Horses

Decision that no MRL is necessary

MRL not necessary for the protection of human health because use of the substance in food-producing animals would not result in residues of concern

Candidates:

- Residues well below ADI at „zero" withdrawal time
    - poor or absent absorption
    - rapidly and extensively detoxified or excreted.
- Substance is of endogenous origin
- Substance is a normal component of the diet in humans
- Substance biologically inert

Classification of Maximum Residue Limits

4 Classes of substances
(Comparable to Annexes 1, 3, 2 and 4 of Regulation 2377/90)

1. A maximum residue limit
    - On the base of opinion of EMEA
    - Vote by Community in Codex Alimentarius

2. A provisional maximum residue limit
    - Incomplete data, but no grounds for supposing a relevant risk

3. The absence of a maximum residue limit
    - Not necessary for protection of human health

4. A prohibition on the administration of a substance
    - Where any residue may constitute a hazard
    - where no final conclusion concerning human health with regard to residues of a substance can be drawn.

Entry of MRL in Regulation

**COMMISSION REGULATION (EC) No …/..
of […] on pharmacologically active substances and their classification regarding maximum residue limits in foodstuffs of animal origin**

Pharmacologically active Substance	Marker residue	Animal Species	MRL	Target Tissues	Other Provisions	Therapeutic Classification
Benzylpenicillin	Benzylpenicillin	All food producing species	50 µg/kg 50 µg/kg 50 µg/kg 50 µg/kg 4 µg/kg	Muscle Fat Liver Kidney Milk	For fin fish the muscle MRL relates to 'muscle and skin in natural proportions'. MRLs for fat, liver and kidney do not apply to fin fish. For porcine and poultry species the fat MRL relates to 'skin and fat in natural proportions'. Not for use in animals from which eggs are produced for human consumption.	Anti-infectious agents/ Antibiotics

Analytical methods

- Setting of MRLs requires availability of analytical methods to monitor residues (no MRL withour analytical method)

- There will be a separate presentation on validation of the analytical methods

RISK MANAGEMENT

Setting of withdrawal periods in the markeing authorisation procedure)

Withdrawal Period - Legislative Background

DIRECTIVE 2001/82/EC (new Directive 2004/28/EC)

A withdrawal period is:

"The _period necessary between the last administration_ of the veterinary medicinal product to animals, under _normal conditions_ of use and in accordance with the provisions of this Directive, and the _production of foodstuffs_ from such animals, in order to _protect public health by ensuring that such foodstuffs do not contain residues in quantities in excess of the maximum residue limits_ for active substances laid down pursuant to Regulation (EEC) No 2377/90"

Withdrawal Period - Legislative Background

Two current ways to obtain a withdrawal period:

1. **Decentralised procedure** (mutual recognition): Applicant seeks approval in a MS. After approval, that MS refers application to other EC countries (Directive 2001/82/EC, as amended by Directive 2004/28/EC)

2. **Centralised procedure**: based on direct evaluation from the EMEA/CVMP leading directly to single marketing authorisation valid throughout the EU (CR 2309/93, as amended by CR 726/2004)

Guidelines and Calculation Programs for Withdrawal Periods in Meat and Milk

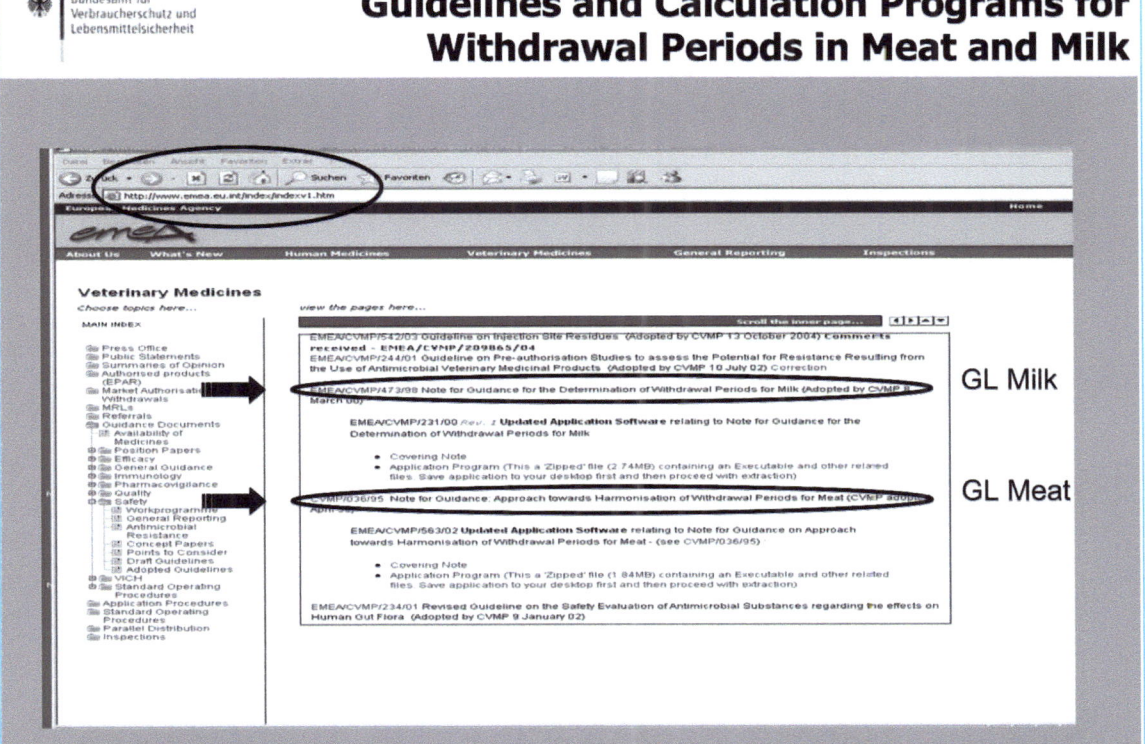

Basic Residue Study Requirements

- Withdrawal periods are determined from results of <u>residue depletion studies</u> in the <u>intended species (target species)</u>,

- and are conducted under <u>simulated field-use conditions</u> with the <u>intended product</u> (same route of administration, dosage and duration of treatment

- Depletion of residues (e.g., marker residue) in meat, milk, eggs etc must be <u>monitored from the time of treatment until (at least) the time when concentrations fall below MRLs</u>

Principles for Establishing Withdrawal Periods

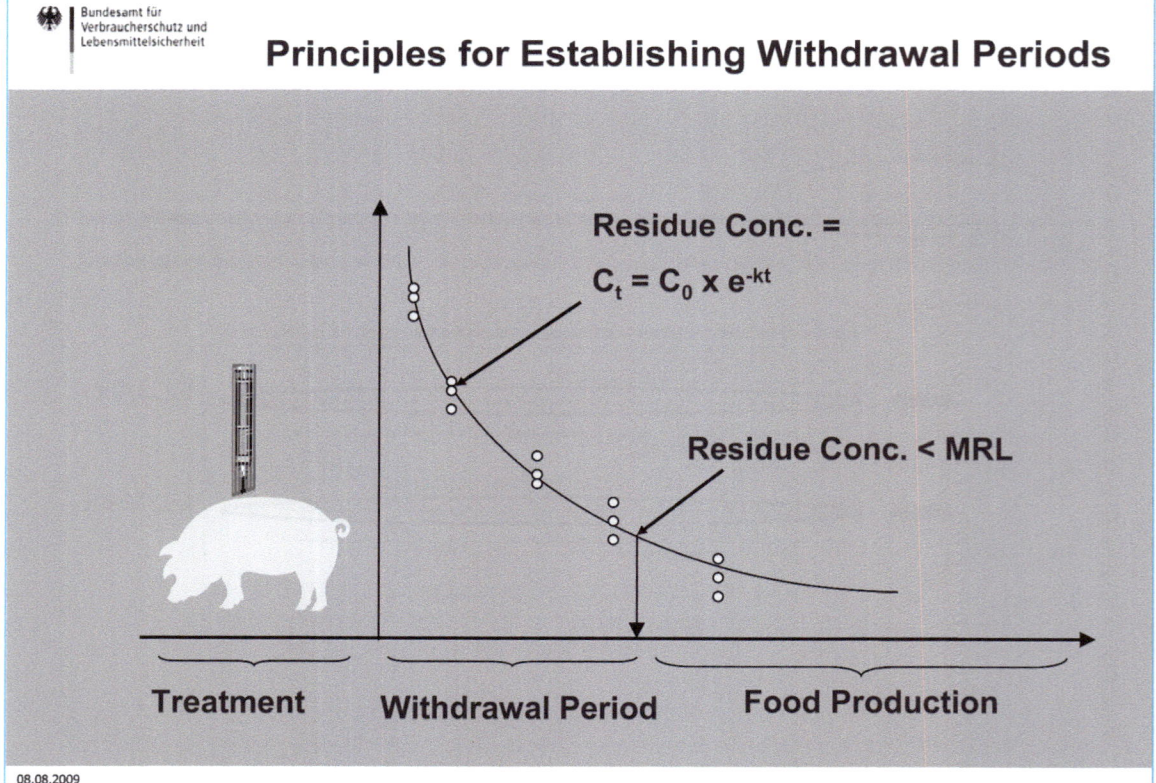

Residue Conc. = $C_t = C_0 \times e^{-kt}$

Residue Conc. < MRL

Treatment | Withdrawal Period | Food Production

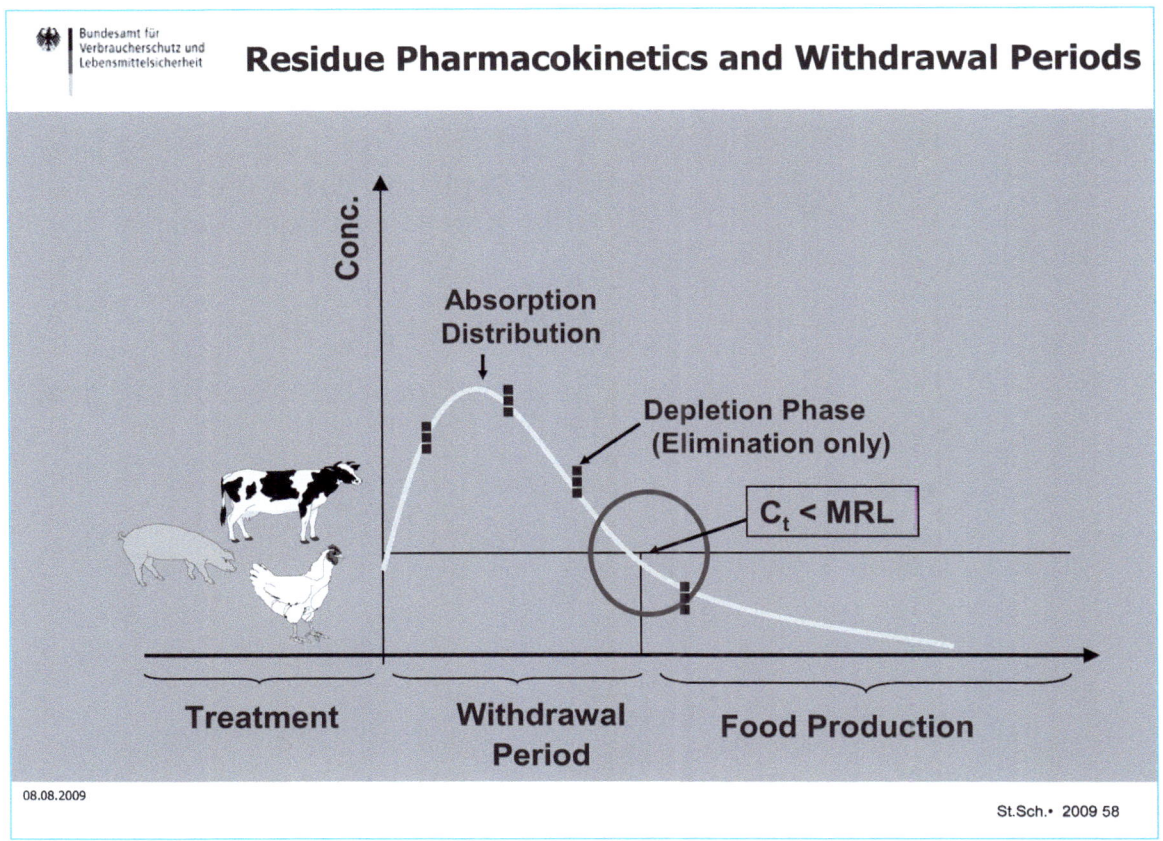
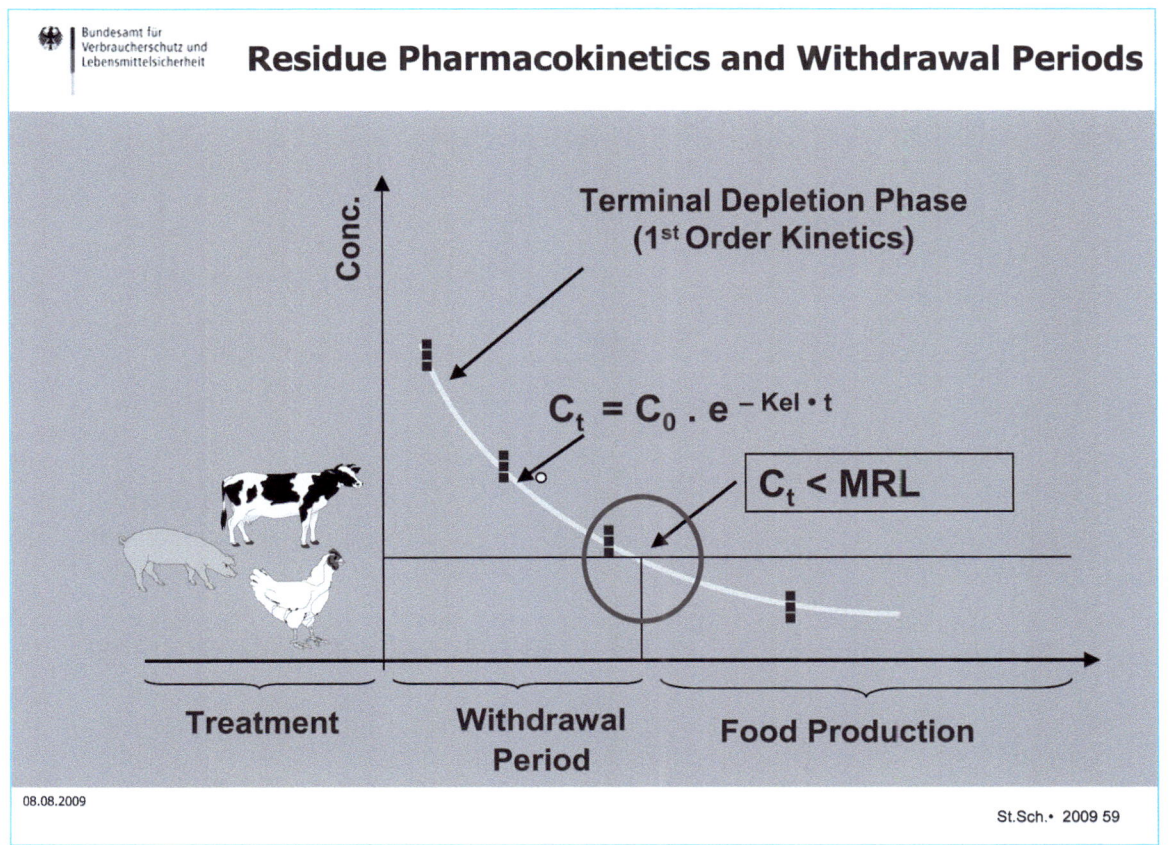

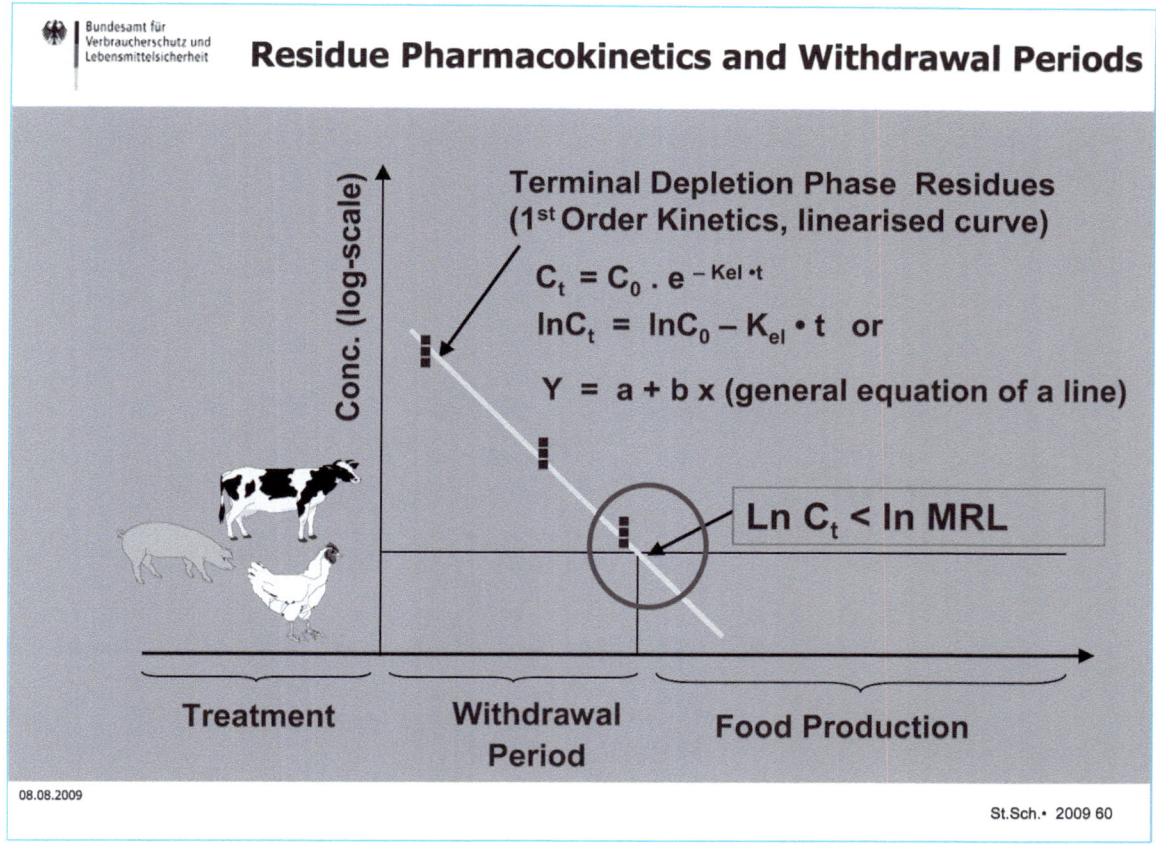

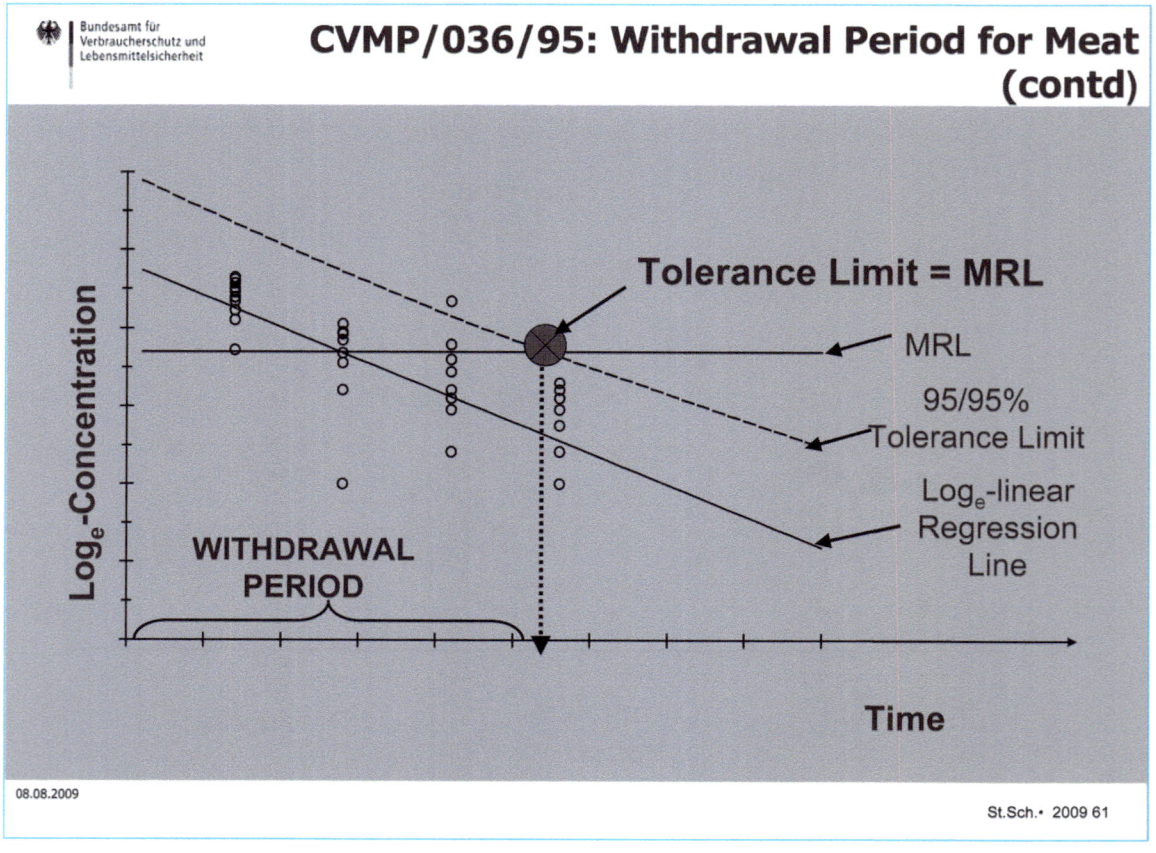

Alternative Approach to establish Withdrawal Periods (all commodities)

- To be used when data do not permit use of statistical approach

- One typical approach to the problem: Selection of time point when concentrations of residues in all samples are below the relevant MRLs

- Addition of a "safety span" to compensate for uncertainties due to biological variability:

  - 10-30% of time period to reach MRL in all sampels
  - for tissues e.g. depletion half-life (1-3 x $t_{1/2}$)

Injection Sites- Specific Properties of Injection Site Residues

- <u>Depletion of residues can be very slow and erratic, and</u> often <u>much higher concentrations</u> than in non-injection site tissues

- <u>Metabolism/degradation of residues</u> at injection sites often different from other edible tissues

- <u>Sometimes wide and uneven dispersion of residues</u> from the point of injection and specific sampling technique is necessary to accurately sample/determine the residues

- Injection site residues <u>tend to be non-compliant with</u> withdrawal periods for other edible tissues

CVMP/542/03: Guideline on Injection Site Residues

- EMEA/CVMP "Guideline on Injection Site Residues (EMEA/CVMP/542/03, October 2004)

- **Basic assessment principle:** The injection site is treated as "normal" muscle and the withdrawal period to be set on the basis of the MRL for muscle (if there is a muscle MRL !)

- For substances with **no MRL for muscle,** the injection site is treated as muscle and the withdrawal period is set with reference to the ADI and the "food basket" (here the 300 g muscle portion is substituted by 300 g injection site)

CVMP/542/03: Guideline on Injection Site Residues

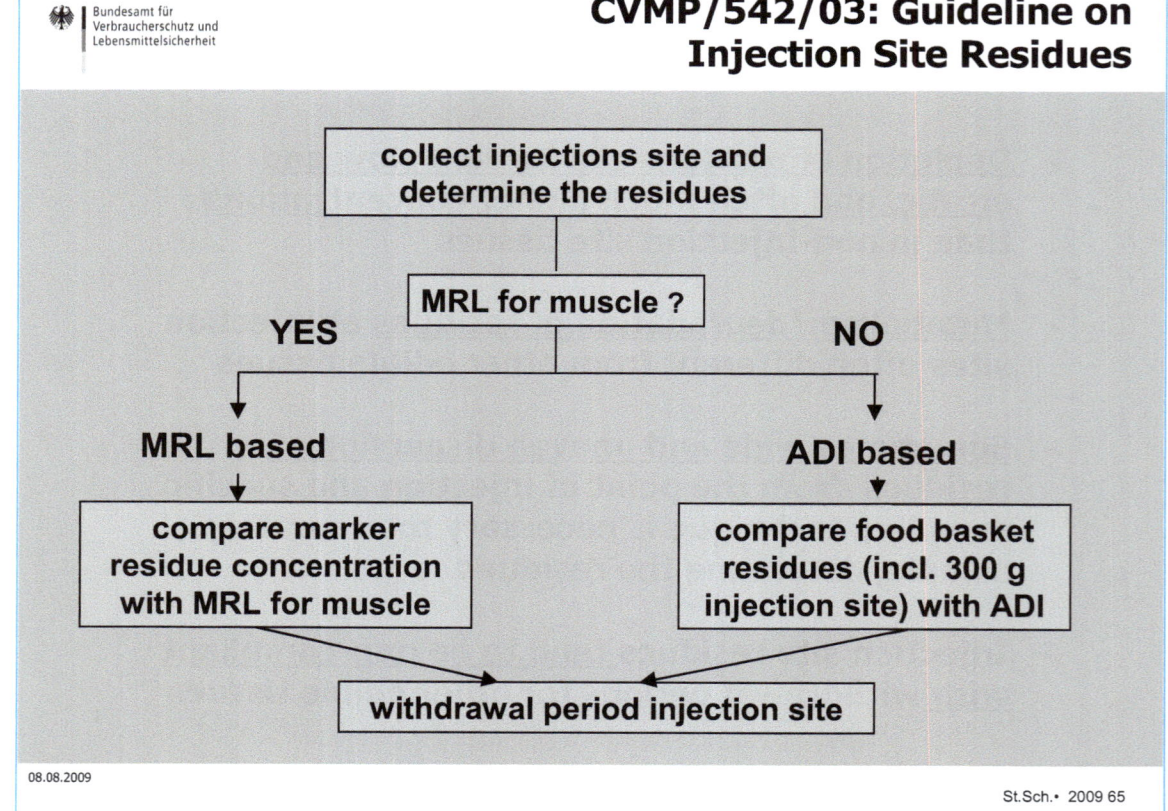

CVMP/542/03: Guideline on Injection Site Residues

Sampling in at injection sites:

1. **Core Sample (Primary Sample)**

   Cylinder-shaped sample taken at the centre of the injection site (incl. the needle track, area of drug release, any area of tissue reaction).
   Sample ca. 500 g amounts (i.m.: 10 cm circle, 6 cm deep, s.c.: 15 cm circle, 2.5 cm deep)

2. **Surrounding Sample (Quality Control Sample)**

   taken from tissue surrounding the core sample to check reliability of sampling. Suggested amount ca. 300 g, if possible

CVMP/542/03: Guideline on Injection Site Residues

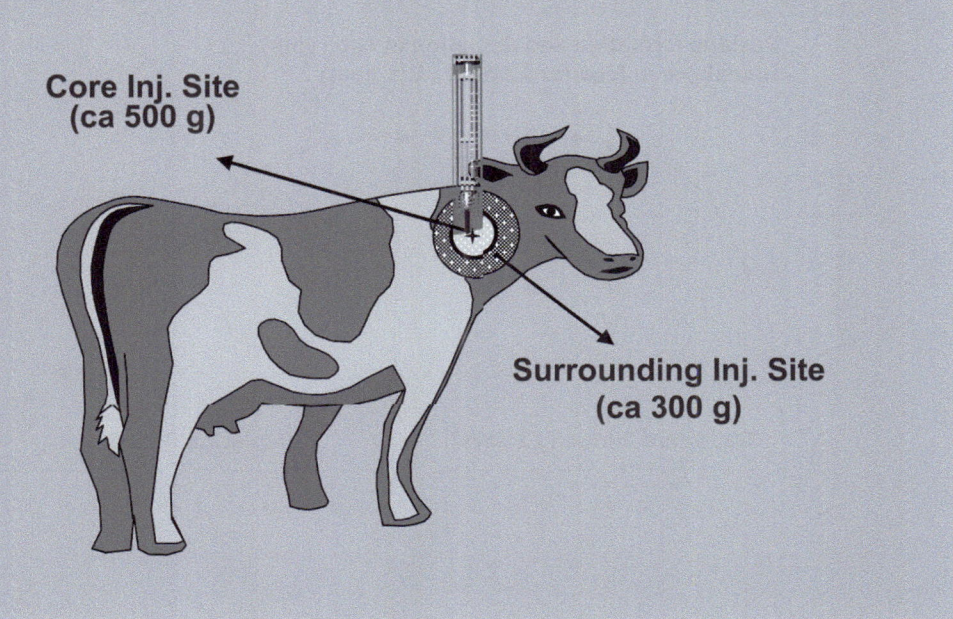

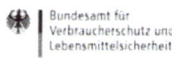 

Specific Aspects for Eggs

Eggs:

- No specific guidelines available, no recommended statistical method

- Withdrawal time must consider physiological egg development (possible lag-phase of depletion after treatment), samples have to be collected over a sufficiently long period (min. 10 days)

- Products <u>without MRL</u> for eggs:

→ <u>laying hens</u> not allowed to be treated !

→ <u>pre-lay hens</u> can be only treated until a definite upper age limit

→ <u>Estimation of upper age limit</u> :

  + set anticipated point-of-lay (20 weeks of age)
  - minus time of the fast egg development (10 days)
  - minus withdrawal period for tissues (e.g. 2 days)
  - minus treatment period (e.g. 5 days)

Specific Aspects for Eggs

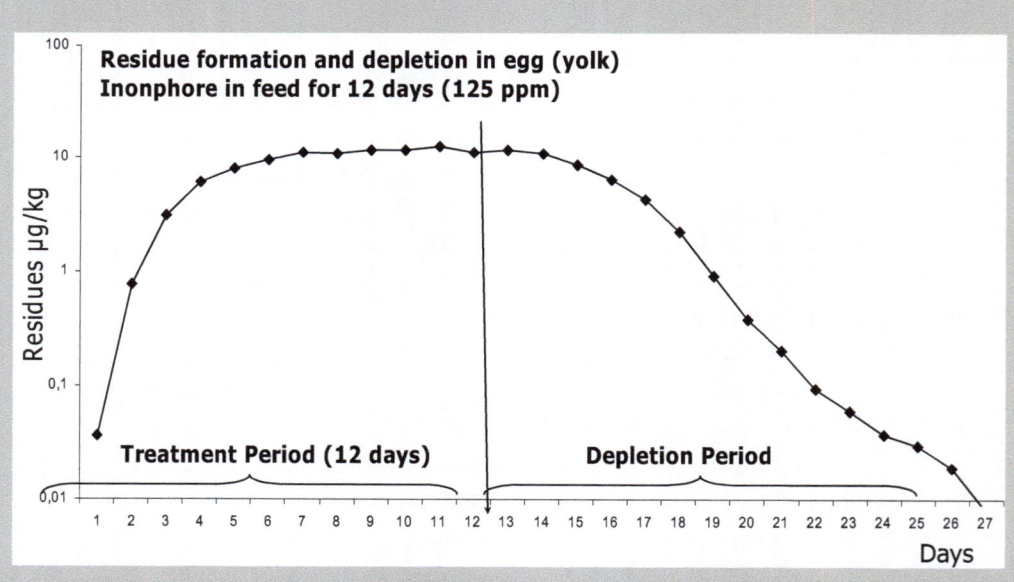

Specific Aspects for Fish

Fish:

- Metabolism/elimination of drug correlated with water temperature.

- Withdrawal period is a function of time <u>and</u> temperature and is expressed in <u>degree days</u>

- Calculated by multiplying the total number of days needed to reach safe concentration by the mean daily water temperature (in degrees Celsius)

- For instance, a withdrawal period of 175 degree-days is corresponding 17.5 days at a water temperature of 10 degrees Celsius

Specific Aspects for Honey

Honey:

- No decrease of residues as a result of pharmacokinetics (residues that go into honey largely remain there)

- Reduction of residues mainly due to <u>dilution effects</u> (function of honey yield) and, possibly, <u>chemical decay</u>

- These variables largely unpredictable (seasonal /geographical differences, etc) and not directly related to a specifiable period of time

- Therefore, the only feasible WP in honey is a "<u>nil</u>" WP which is <u>supported by a range of suitable studies</u> demonstrating that residues are below MRLs under various/variable conditions of good bee keeping practice

Analytical methods

- Setting of MRLs and establishment of withdrawal periods requires availability of an analytical methods to monitor residues (no MRL withour analytical method)

- Basic validation criteria for methods are presented hereafter

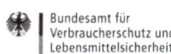

Requirements for Analytical Methods

It is not possible to set maximum residue limits without an analytical control method!!!

"Volume 8":

- ✓ "Applicants should [...] make available a validated analytical method, which can serve as a basis for official <u>residue monitoring and</u> surveillance ("regulatory method")"...

- ✓ "This method should determine the marker residue on which the MRLs are based"

- ✓ A method is required for all target tissues (incl. milk, eggs, honey, if applicable)

Requirements for Analytical Methods - Validation

- Method <u>validation performed by applicants</u> seeking establishment of MRLs; the work is done in their own labs/contracts labs

- <u>EMEA performing a "desk review"</u> of validation parameters, to assess its suitability for detecting/quantifying residues at concentration of interest (MRL)

- <u>Methods are made available to Community reference laboratories (CRLs)</u> and national reference laboratories where they are put into practice, re-validated, further adapted to meet the needs of these laboratories

Requirements for Analytical Methods - Validation

Validation criteria according to „Volume 8"

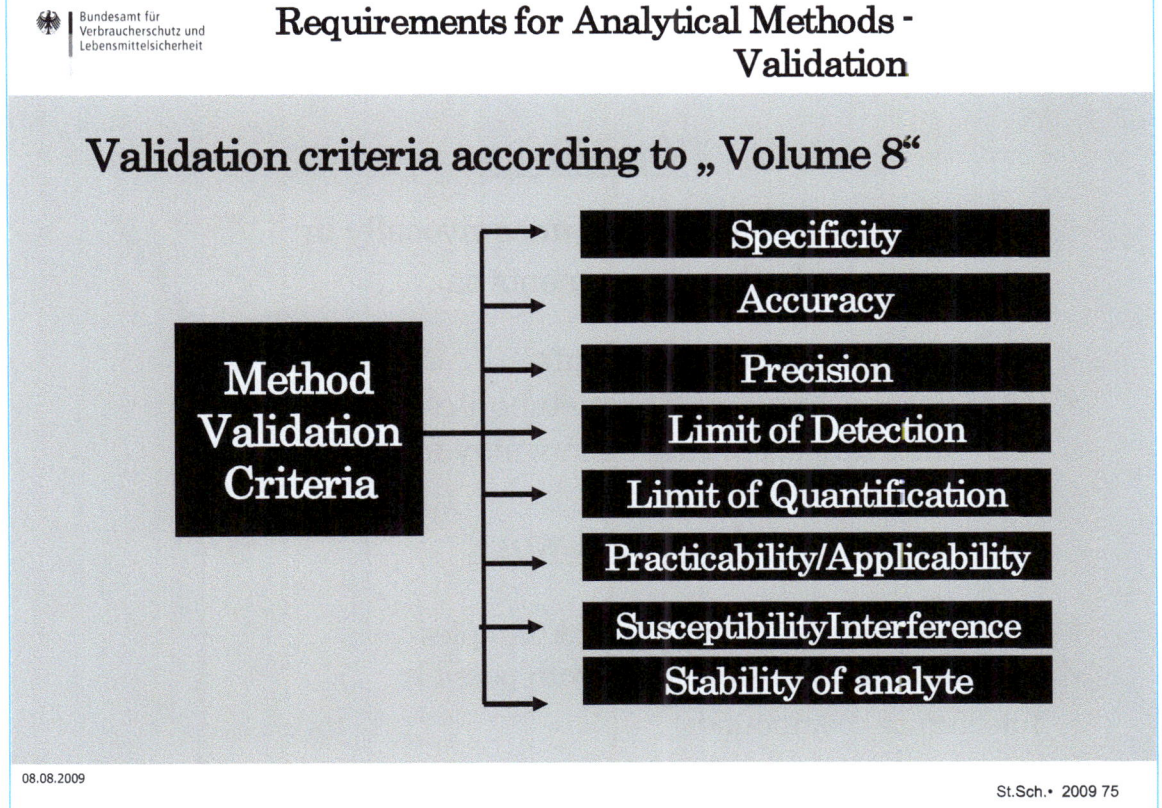

Method Validation Criteria:
- Specificity
- Accuracy
- Precision
- Limit of Detection
- Limit of Quantification
- Practicability/Applicability
- Susceptibility Interference
- Stability of analyte

Requirements for Analytical Methods - Validation

Test materials

1. Blank tissues „spiked" with the test substance

   OR

2. Certified reference materials (if available),

   OR

3. Incurred tissues from treated animals (if applicable)

Requirements for Analytical Methods - Specificity

Specificity

Ability to analyse an analyte unequivocally in presence of other components as......

- ✓ chemically related components
- ✓ homologues, analogues or metabolites,
- ✓ also naturally occurring agents may interfere

Practical work

- ✓ Representative number of blank samples
- ✓ also blank samples «spiked» with possibly interfering compounds

Requirements for Analytical Methods - Accuracy

Accuracy (1)

Agreement between measured value and value accepted either as "true value" or a "accepted reference value"

Principal limitations:

- ✓ Systematic errors (bias)
- ✓ Random errors (i.e. occational low recovery);

Requirements for Analytical Methods - Accuracy

Accuracy (2)

Accuracy expressed as:

- ✓ "Recovery" ("spiked" standards)
- ✓ "Trueness" (certified reference materials)

Practical work

18 mutually independant samples fortified at:
- 6 x 0.5 x MRL
- 6 x 1 x MRL
- 6 x 2 x MRL

Requirements for Analytical Methods - Accuracy

Accuracy (3)

Calculated as:

$$\frac{\text{Measured Value}}{\text{Target Value}}$$

Accepted Limits

Content (Mass fraction)	Limits
< 1 µg/kg	0.5 and 1.2
> 1 µg/kg	0.7 and 1.1 (range for most MRLs!)

For large numbers of measurements the accuracy approaches the systematic error!!!

Requirements for Analytical Methods - Precision

Precision

Closeness of agreement between repeated analysis - measure of the dispersion of random errors

Repeatability:
intra-assay or intra-day variability (within a short time)

Within laboratory reproducibility:
inter-assay or between-day variability (in one and the same laboratory)

Not necessary for MRL applications!!!
Reproducibility: precision between laboratories (usually applied to standardization of methodology)

Requirements for Analytical Methods - Precision

Precision - Repeatability

A measure for repeatability is the relative standard deviation or co-efficient of variation (CV%)

Practical work

3 series of recovery analyses (n=6/level) incl. 0.5x to 2x MRL

Accepted Limits

Content (Mass fraction)	CV%
< 1 µg/kg	35%
> 1 µg/kg and < 10 µg/kg	30%
> 10 µg/kg and < 100 µg/kg	20%
> 100 µg/kg	15%

Requirements for Analytical Methods - Precision

Precision - Within laboratory reproducibility (1)

Within-laboratory reproducibility CVs should not exceed levels calculated by Horwitz equation $CV = 2^{(1 - 0.5 \log C)}$

Practical work

3 series of recovery analyses (n=6/level) incl. 0.5x to 2x MRL (different operators, different environmental conditions, …)

Requirements for Analytical Methods - Precision

Accepted Limits

Representative within-laboratory CVs for quantitative assays are:

Content (Mass fraction)	CV %
100 µg/kg	<23
1000 µg/kg	<16

<u>Rule:</u> for concentrations lower than 100 µg/kg, the CVs should be as low as practicable possible without exceeding the CVs given by the Horwitz equation

Requirements for Analytical Methods - Accuracy & Precision

Validation - Relationship between accuracy and precision

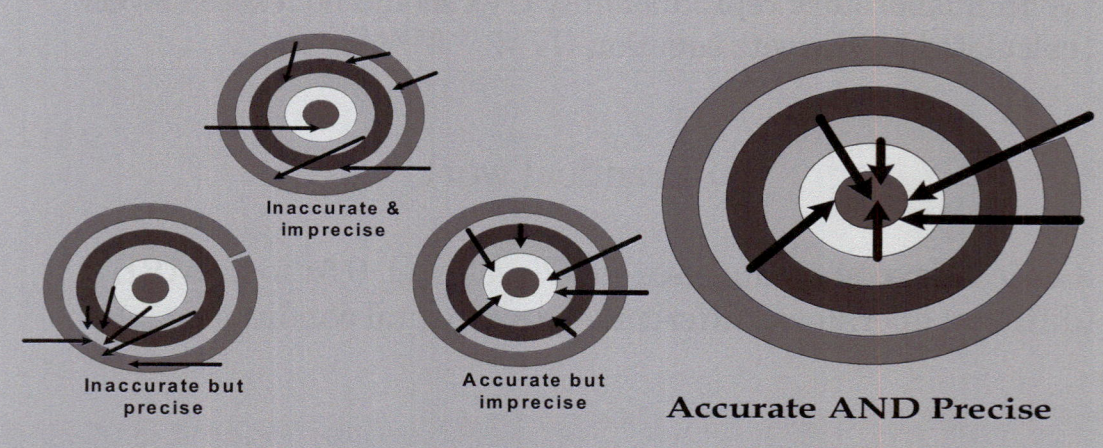

Requirements for Analytical Methods - Limit of Detection

Limit of Detection (LOD)

Lowest concentration that can be distinguished from background (but not necessarily quantified)

Practical Work

- Arithmetic mean of concentration in blank samples (n = 20) plus three times the standard deviation
- Other methods of calculation providing substantially the same statistical certainty may be used

---

Requirements for Analytical Methods - Limit of Quantification

Limit of Quantification (LOQ)

The limit of quantification should not be greater than half the MRL (the lowest limit to be „fully" validated is ½ x MRL)

Accepted Limits

The limit of quantification is the lowest amount of analyte in a sample which can be quantitatively determined with suitable precision and accuracy

Requirements for Analytical Methods - LOQ/LOD

Validation – LOD/LOQ & Background

$LOQ = \text{validated} \leq \tfrac{1}{2} MRL$

$LOD = \text{mean BG} + 3\, SD$

Background noise

Requirements for Analytical Methods - Stability

Stability of the Analyte

should be tested:

- in solution during analysis/storage (e.g., working solutions)
- in matrix during sample preparation/storage
- in extract during storage/analysis

Practical Work

Example: Analyte in matrix: fortified blank samples at level of interest under storage conditions (- 20 °C)
Analysis at time 0, after 1 week, 2 weeks, … as long as required (e.g. 20 weeks)

Requirements for Analytical Methods - Susceptibility

Susceptibility to Interference

Any variation, which could affect the analytical results

……..non-specific influences on the analytical results of certain experimental conditions (e.g. stability of reagents, pH, temperature, composition of the matrix etc)

Requirements for Analytical Methods - Practicability & Applicability

Practicability & Applicability

Practicability:
Refers to the scope of the method and is determined by requirements, such as availability of standards, reagents and equipment, sample through put/time or costs.

Applicability:
Refers to the species, the matrices (e. g. liver), matrix conditions, concentration range etc for which method can be applied as described or with minor modifications

Applicability should be demonstrated for all matrices for which MRLs exist

Requirements for Analytical Methods – Calibration curve

Calibration curve

A typically calibration curve:

- at least 5 equaly spaced concentrations (incl. «0»)
- Description of working range
  - Determination of the mathematical model, goodness of fit, intercept/slope

Practical work

5 levels of concentrations: 0 – 0.5 – 1 – 1.5 – 2 x MRL

Summary of Validation Tools

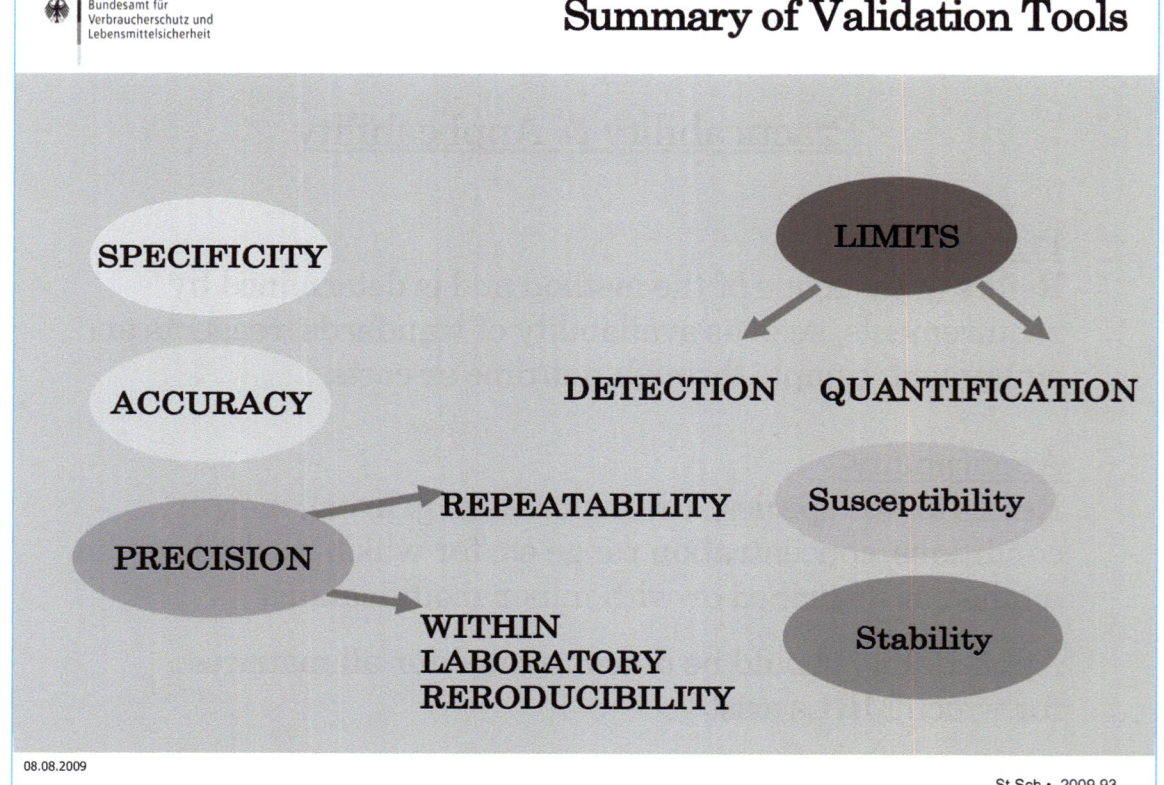

Frequently used methods

Methods need to meet criteria for confirmatory methods

Typical methodoligies:

LC or GC with mass-spectrometric detection
LC-fluorescence
LC-DAD
LC-UV/VIS

Development of screening method not required
Same criteria for methods used in residue trials !!!!

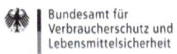

COMMISSION DECISION
of 12 August 2002
implementing Council Directive 96/23/EC concerning the performance of analytical methods and the interpretation of results
(notified under document number C(2002) 3044)
(Text with EEA relevance)
(2002/657/EC)

Lays down rules for   identification
                     validation
                     interpretation
of analytical methods for drug residues
MRL and banned compounds

User Safety

EMEA/CVMP Draft Guideline on

User Safety for Veterinary Medicinal Products

Legal Basis

Directive 2001/82: *"safety documentation shall show the potential risks which may result from the exposure of human beings to the medicinal product, for example during its administration to the animal"* and *"the dossier shall inculde a thorough discussion of any risks for persons preparing the medicinal product or administering it to animals, followed by proposals for appropriate measures to reduce such risks"*

Since 13 July 2005: CVMP Guideline on User Safety

General Principles

- User safety assessment is required for <u>all veterinary products</u> to be marketed

- User safety is assessed in the <u>context of granting marketing authorisation</u>

- User safety is <u>product specific</u> and considers active ingredient(s) and all other excipients present in the formulation.

- User safety relates to <u>all persons</u> in charge of treating the animals and those handling the products and treated animals (veterinary surgeons/ nurses, farmers/farm workers, animal handlers, pet owners and children etc)

General Principles (contd.)

- User safety assessment involves two major assessment steps (1) <u>identification/ characterisation of relevant hazards</u> and <u>(2) the assessment of likelihood of exposure</u> from administration of the product and contact with treated animals

- User safety is <u>generally managed by</u> <u>avoidance of unnecessary exposure</u> with advice on safe storage, handling and disposal of the product (warning phrases in the product literature should prevent risks under foreseeable circumstances of use)

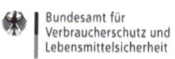

Data requirements

User safety assessment is mainly based on scientific information compiled in Part 3 A (Safety Documentation) of the marketing application dossier

Sections of the application dossier that are particularly relevant to user safety are:

> *Pharmacokinetics (ADME)*

> *General toxicology (type and severity of hazards)*

> *Single dose toxicity* (hazards of final product by e.g. dermal or inhalation exposure,

> *Reproductive toxicity* (significance of test results in view of user safety)

Toxicity Data

The toxicity data should provide information on the possible adverse effects with a view to likely exposure scenarios:

> Both local and systemic effects need to be taken into account

> The <u>systemic</u> effects for <u>active ingredients</u>

> For local toxicity, the formulated test article(s)

> The dose-response relationships (NOAEL, LOAEL)

> Toxicity studies should ideally use the same routes of exposure and duration as in the anticipated exposure scenarios

Minimum Toxicity Data Requirements

Minimal required toxicity data according to EMEA/CVMP/543/03

Relevant route of exposure	Minimal required information regarding:	
	for the whole product (active ingredients plus excipients)	for active ingredients
Oral	Acute oral toxicity	All relevant toxicity data and human data (when available) in accordance with the requirements for dossier parts III.A.1 through III.A.4
Dermal	Skin irritation Skin sensitisation Photo toxicity	
Parenteral (i.e. self-injection)	Acute parenteral toxicity (local/systemic)	
Inhalatory	Respiratory irritation Respiratory sensitisation	
Ocular	eye irritation[1]	eye irritation[1]

1) If the test article is irritating to the skin, it is assumed that it is also irritating to the eyes and no eye-irritation test has to be performed for skin irritating test articles.

Exposure Scenarios

The exposure scenario comprises a description of:

➢ the type of user

➢ the routes of exposure

➢ the components of a product to which the user is exposed

➢ the likelihood of exposure

➢ The type, rate, extent, duration, interval, and frequency of exposure

Management Options

The guideline differentiates between non-professional and professional users:

1. <u>Non-professional users</u> (person in the general public/consumers can come in to contact with the VMP):

   The risk should be acceptable for unprotected non-professionals, or with only very limited protection

2. <u>Professional users</u> (eg, veterinarian, farmer, breeder etc):

   When there is an unacceptable high risk for the unprotected professional user, measures to reduce the risk to an acceptable level can be proposed

Summary of Assessment Approach

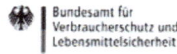

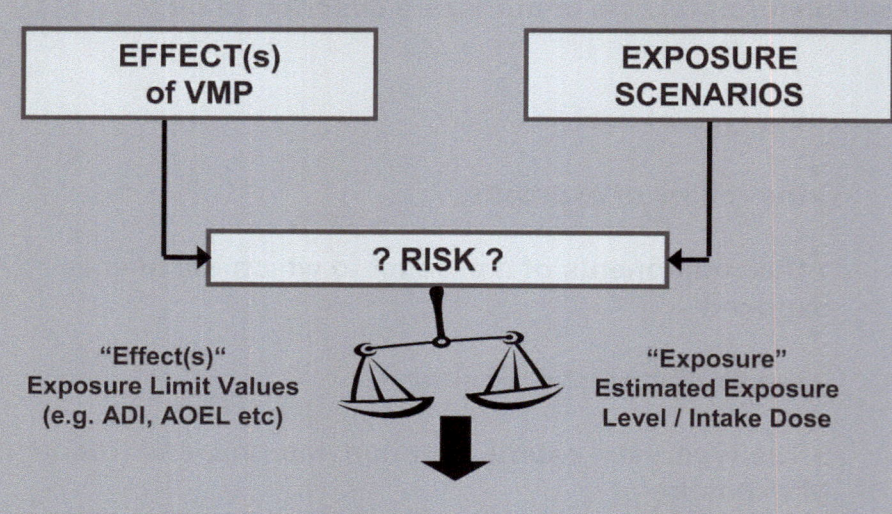

2 Target Animal Safety for Veterinary Pharmaceutical Products (VICH GL 43)

Gesine Hahn

Federal Office of Consumer Protection and Food Safety (BVL), Berlin

Correspondence to: Dr. G. Hahn, BVL, Ref. 303, Mauerstraße 39–42, D-10117 Berlin, Germany, Tel.: 030 18 444 30300, e-mail: gesine.hahn@bvl.bund.de

Presentation

- History
- Process
- Key issues
- Conclusions and outlook

Gesine Hahn • 8. August 2009

How did it start?

EU: Evaluation of TAS, 1994

Local tolerance of intramam. products, 1993

USA: TAS guideline for new animal drugs, 1989

TAS and drug effectiveness studies for antimicrobial mastitis products, 1996,

Protocol for clinical effectiveness and TAS trials, 2001

JP: Guidelines for Toxicity Studies for registration of new animal drugs, 1988

Gesine Hahn• 8. August 2009 •

How did it start?

- Differences in the requirements for TAS testing identified:
  - Extensive use of pharm-tox background information (EU)
  - Drug tolerance testing and toxicity testing in the target species (FDA)
  - Study protocols, extensive list of items to be tested (FDA)
  - Special studies, e.g. reproductive safety (JP, U.S.)

Gesine Hahn• 8. August 2009 •

How and when did it start?

Mandate from VICH SC:
- Develop internationally harmonized guidance for TAS studies for veterinary products incl. pharmaceuticals and biologicals*)
- Agree on the minimum requirements for the safety data package with an objective to assure well being of the target animal
- Species-specific/ formulation-specific guidelines, if considered necessary

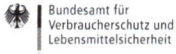

The process

- VICH WG started work in Nov 2000
- WG consensus achieved in Sept 2006 (step 2)
- CVMP adoption for release for consultation in Nov 2006 (step 4)
- WG consensus in June 2008 (step 5)
- CVMP approval in Sept 2008 (step 7)
- Date coming into effect July 2009

Table of content

- INTRODUCTION
  - Objective
  - Background
  - Scope
- MARGIN OF SAFETY STUDIES
  - Standards
  - Animals
  - IVPP and Route of Administration
  - Dose, Frequency, and Duration of Administration
  - Study Design
  - Variables
  - Statistical Analysis
- Study Reports
- OTHER LABORATORY SAFETY STUDY DESIGNS
  - Injection Site Safety Studies
  - Administration Site Safety Studies for Dermally Applied Topical Product
  - Reproductive Safety Studies
  - Mammary Gland Safety Studies
- TARGET ANIMAL SAFETY DATA FROM FIELD STUDIES
- RISK ASSESSMENT IN ANIMAL SAFETY EVALUATION
- GLOSSARY

Gesine Hahn• 8. August 2009

Comparison to EU TAS guideline

- Much longer
- Information given in more detail
- Exhaustive list of variables
- Special studies (reproductive safety, mammary gland safety)
- Flexibility maintained

Gesine Hahn• 8. August 2009

Aim of TAS studies

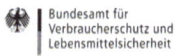

- To provide information on the safety of the IVPP in the intended species under the proposed conditions of use
- To establish a margin of safety unless exposure is low
- To identify adverse effects associated with overdoses and/or increased duration of use, if possible

Scope

<u>Animal species</u>
- Cattle, sheep, goat, swine, horses
- Cats, dogs
- Poultry (chicken, turkeys)

For other species, national or regional guidance should be followed.

Margin of Safety study

Standard:
- In conformity with the principles of GLP
- Concept of GMP as appropriate for new animal drugs

Animals:
- Healthy young mature animals unless the product is for use in young immature animals;
- Additional studies in sensitive subpopulations, if identified
- Animal welfare consideration is essential

Margin of Safety study

IVPP and route of administration:
- Commercial formulation
- Comparative (bridging) studies, if necessary
- Dosing should follow the proposed use conditions
- If multiple routes are proposed, the one most likely to cause adverse effects should be selected

Margin of Safety study

Dose levels:

Unless otherwise justified by pharm-tox properties of the API and the proposed use of the product

– 0, 1X and two multiples of the use dose (in most cases 3X and 5X*)

(1X defined as the highest dose that will be stated in the product literature)

*Recommendation based on a scientific survey in the VICH regions EU, U.S. and Japan

Margin of Safety study

Frequency and duration:

Unless otherwise justified at least 3 times the recommended duration of use up to max. 90 days

- single use: 3 consecutive intervals as determined by the pharmacologic characteristics of the product
- daily use: 3X the proposed duration
- intermittent use: 3 treatments at the recommended interval
- For products intended for use longer than 3 months, duration up to 6 months or longer, if appropriate

Margin of Safety study

Study design:
- Randomisation and blinding most important to avoid bias
- At least 8 animals/ group (4 f, 4 m)
- Potential impact of group housing to be considered
- Personnel collecting data including gross post mortem results should be masked

Margin of Safety study

Variables:
- Physical examinations and observations
- Clinical pathology tests (to be selected depending on the nature of the IVPP and the intended population)
- Necropsy and histopathology
  - New API: All animals in all dose groups
  - Others: As a minimum, animals of the highest and control group; determine NOAEL, if lesions are found
    Alternative schedule: subsets of animals selected at random prior to the study
    IVPPs with well-documented broad margin of safety may be dispensed from post mortem exam.

Margin of Safety study

Statistics:
- Use of descriptive statistical methods
- Presentation of results in text, tables and/or graphs
- Statistical models should be selected based upon the study design and the nature of response variables
- Statement of significance levels is recommended
- Clinical significance of statistically (non-) significant results should be considered

Gesine Hahn• 8. August 2009 •

Injection Site Safety studies

- 1X dose, duration, route and max. volume
- Saline control at the same volume
- Appr. time required to return to clinically acceptable conditions
- IV route – extravenous administration to be considered
- Clinical signs, signs of local inflammation
- CK and AST, histopathology, if clinical signs indicative of injection site reactions

Gesine Hahn• 8. August 2009 •

Administration Site Safety studies for topical products

- Dosage as proposed
- Appr. time required to return to clinically acceptable conditions
- Clinical signs, local signs of inflammation
- Histopathology, if clinical signs indicative of application site reactions
- Oral dosing study at the max. proposed dose if accidental ingestion is likely to occur

Reproductive safety studies

- To identify adverse effects on male/ female reproduction performance and viability of offspring
- At least 8 healthy intact animals per sex per treatment
- Usually 0 and 3X dose
- Data derived from lab animals may be considered, if pharmacokinetic profiles are comparable
- The absence of studies in the target species should be reflected in the product literature

Mammary gland safety studies

IVPPs intended for lactating/ non-lactating animals:

- Evaluation of acute inflammatory effects (GLP)
    - lactating animals in early to mid-lactation
    - 1X dose/quarter, frequency and duration as proposed
    - Parameters: Physical examination, QSCC, bacterial culture
      Milk yield, composition and appearance
    - One group design / Parallel group design

- Evaluation of chronic inflammatory effects of IVPPs for non-lactating animals (GLP/ GCP)

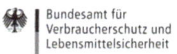

TAS data from field trials

- Conducted in accordance with GCP
- Evaluate potential adverse effects at the intended dosage regime in a broad target population
- International data may be used where disease and husbandry conditions are similar
- Observation and causality determination of adverse events

Conclusion and outlook

- Consensus document considered as a balance of the different views of the agencies
- Minimum requirements for TAS evaluation
- Should minimize animal testing
- Should avoid duplication of similar studies in each studies
- Animal welfare should benefit

Gesine Hahn • 8. August 2009 •

3 Beurteilung und Überwachung der Resistenzsituation bei und nach der Zulassung von Tierarzneimitteln

Christine Schwarz

Bundesamt für Verbraucherschutz und Lebensmittelsicherheit (BVL), Berlin

Korrespondenz an: Dr. C. Schwarz, BVL, Ref. 303, Mauerstraße 39–42, 10117 Berlin,
Tel.: 030 18 444 30318, e-mail: christine.schwarz@bvl.bund.de

---

Tierarzneimittelzulassung
AMG, RL 2004/28/EG zur Änderung der RL 2001/82/EG

Bei der Zulassung von Antibiotika bestehen grundsätzlich gleiche Anforderungen für neue und bekannte Substanzen, die Wirksamkeit ist durch klinische Studien zu belegen, es gibt keine spezifischen Regelungen hinsichtlich MRSA

- Für alle Tierarten: Resistenzgutachten
  - Guideline for the demonstration of efficacy for veterinary medicinal products containing antimicrobial substances (EMEA/CVMP/627/01-FINAL)
    - Repräsentative Daten zur Empfindlichkeit von Zielerregern aus den letzten 5 Jahren (MHK50, MHK90), Pharmakodynamische Daten, Daten zur Resistenz (Entwicklung, Mechanismen, Co/Kreuzresistenz)

- Bei Lebensmittel Liefernden Tieren (LLT): zusätzliches Resistenzgutachten
  - Guidance on Pre-Approval Information for Registration of Veterinary Medicinal Products for Food Producing Animals with Respect to Antimicrobial Resistance. VICH Topic GL 27(EMEA/VICH/644/01-FINAL)
    - MHK Daten zu Zoonoseerregern (*Salmonella enterica*, *Campylobacter* spp.) und zu
      Kommensalen (*Escherichia coli*, *Enterococcus* spp.), weitergehende Unterlagen

Zulassung erfolgt aufgrund der Datensituation

– Antragsgemäß, mit Einschränkungen oder besonderen Hinweisen in der Fachinformation
  - Guideline on the SPC for Antimicrobial Products (EMEA/CVMP/612/01-FINAL)
  - Reflection paper on the use of fluoroquinolones in food producing animals- Precautions for use in the SPC regarding prudent use guidance (EMEA/CVMP/416168/2006-FINAL)

– Auflagen
  - Reflection paper on antimicrobial resistance surveillance as post-marketing authorisation commitment (EMEA/CVMP/SAGAM/428938/2007-CONSULTATION)

Post Marketing – Pharmakovigilanz
AMG, RL 2004/28/EG zur Änderung der RL 2001/81/EG

Die Beurteilung erfolgt produktbezogen, keine gibt keine spezifischen Anforderungen für Antibiotika

– Verlängerung der Zulassung nach 5 Jahren: Resistenzgutachten

– Pharmakovigilanz: PSUR (Periodic Safety Update Reports 4 x halbjährlich, 2 x jährlich, dann alle drei Jahre): Daten zu UAW, einschließlich mangelnder Wirksamkeit
  - VOLUME 9B of The Rules Governing Medicinal Products in the European Union - Guidelines on Pharmacovigilance for Medicinal Products for Veterinary Use
  - VICH Topic GL29 Guideline on pharmacovigilance of veterinary medicinal products management of periodic summary update reports (PSUs) (EMEA/CVMP/VICH/646/01)
  - demnächst: VICH Topic GL24 Guideline on pharmacovigilance of veterinary medicinal products management of adverse event reports (AERs)(EMEA/CVMP/VICH/547/00)

Konsequenzen, die sich aufgrund der Datensituation aus dem Post-marketing ergeben können:

— Einschränkungen in der Fachinformation

— Auflagenerteilung

— Stufenplanverfahren bei begründetem Verdacht

Reflection paper on MRSA in food producing and companion animals in the European Union: epidemiology and control options for human and animal health (EMEA/CVMP/SAGAM/68290/2009)

daraus ergeben sich folgende Empfehlungen der EMEA hinsichtlich der Anwendung von Antibiotika bei Tieren:

— *Prudent use* of antimicrobials as set out by the Codex Alimentarius, the World Organisation for Animal Health (OIE) and the Federation of Veterinarians of Europe, and as discussed in the CVMP strategy on antimicrobials 2006-2010, remains a key measure.

— *Consumption* of antimicrobials in the EU should be monitored to identify any sources of unnecessary use and to target specific counter actions, as required.

— Development of *non-antimicrobial control measures*, such as sanitary measures, for infected or colonised animals should be encouraged.

— Use of *last-resort medicines* for MRSA treatment in men should be avoided in animals.

— When evaluating a marketing authorisation application for a veterinary medicine containing molecules used as last-resort treatment for MRSA infections in man, the CVMP will pay special attention in its assessment of the benefits and risk to the need to ensure the *continued efficacy* of such molecules in human medicine.

If you have any concerns about our products,
you can contact us on
ProductSafety@springernature.com

In case Publisher is established outside the EU,
the EU authorized representative is:
**Springer Nature Customer Service Center GmbH
Europaplatz 3, 69115 Heidelberg, Germany**

Printed by Libri Plureos GmbH
in Hamburg, Germany